J
595
.7
HOV

HA. 13.95
1-3- 2001

Hogansville Public Library

D1456142

I Wonder What It's Like to Be a Grasshopper

Erin M. Hovanec

HOGANSVILLE PUBLIC LIBRARY
600 E. MAIN STREET
HOGANSVILLE, GA. 30230

The Rosen Publishing Group's
PowerKids Press™
New York

To Dad, lots of thanks and much love. Your B.

Published in 2000 by The Rosen Publishing Group, Inc.
29 East 21st Street, New York, NY 10010

Copyright © 2000 by The Rosen Publishing Group, Inc.

All rights reserved. No part of this book may be reproduced in any form without permission in writing from the publisher, except by a reviewer.

Photo Credits: p. 4 © Animals, Animals/Arthur Gloor; p. 7 © James Gerholdt/Peter Arnold, Inc.; p. 8 © Johann Schumacher/Peter Arnold, Inc., © C.Allan Morgan/Peter Arnold, Inc.; p. 11 © Hans Pfletschinger/Peter Arnold, Inc.; p. 12 © Animals, Animals/Ken G. Preston-Mafham, © James Gerholdt/Peter Arnold, Inc.; p.15 © Earth Scenes/Mark Stouffer, © FPG/Ulf Sjostedt,© FPG/Don Hebert; p. 16 © Animals, Animals/T.Jackson/O.S.F., © Matt Meadows/Peter Arnold, Inc., © Animals,Animals/Ken G. Preston-Mafham; p. 19 © John R. MacGregor/Peter Arnold, Inc., © CORBIS, © David Cavagnaro/Peter Arnold, Inc.; p. 20 © C. Allan Morgan/Peter Arnold, Inc.; p. 22 © Gunter Ziesler/Peter Arnold, Inc.

Photo Illustrations by Thaddeus Harden

First Edition

Book Design: Felicity Erwin

Hovanec, Erin M.
    I wonder what it's like to be a grasshopper / by Erin Hovanec.
        p.   cm. — (The Life science wonder series)
    Includes index.
    Summary: Introduces the physical characteristics, habits, and behavior of grasshoppers.
    ISBN 0-8239-5452-8
    1. Grasshoppers Juvenile literature. [1. Grasshoppers.] I. Title. II. Series:
    Hovanec, Erin M. Life science wonder series.
    QL508.A2H68 1999
    595.7'26—dc21

                                                                              99-29649
                                                                                 CIP

Manufactured in the United States of America

Contents

Just Hoppin' By

Have you ever wondered what it's like to be a grasshopper, hopping through meadows and leaping through the air? There's more to a grasshopper's life than jumping through the grass. Did you know that grasshoppers can sing beautiful songs? Did you know that they don't just live in the grass? You can also find grasshoppers in tall trees, on the dusty ground, and in sandy beaches. Did you know that grasshoppers can travel hundreds of miles by jumping and flying?

◀ *Grasshoppers do a lot more than hop around.*

5

Big, Big Bugs

Can you imagine trying to ride a bike if you had six legs? Grasshoppers have six legs, and they use all of their legs to walk. Grasshoppers also have five eyes, one mouth, and a pair of **antennae**, or feelers, that act like a nose. They use their antennae to smell scents. Most grasshoppers have two pairs of wings, but some have only one pair, and a few have none at all.

Grasshoppers are some of the biggest **insects** around. They can be from one to five inches long. Full-grown grasshoppers can even be as long as your whole hand!

6

It might be hard for you to get around on six legs, but for a grasshopper, it's easy. ▶

Long-horned

Short-horned

8

Antennae Long and Short

What if your nose, hands, and ears were all rolled into one? That's what it would be like if you were a grasshopper. Grasshoppers have antennae that they use to smell, touch, and even hear!

Scientists **classify** grasshoppers based on their antennae. Short-horned grasshoppers have short, thick antennae. Long-horned grasshoppers have long, thin antennae that look like threads. These skinny antennae can grow longer than the grasshoppers' bodies. That would be like you having a nose that was bigger than your body!

◀ *The short-horned Mexican grasshopper and long-horned katydid both use their antennae to smell, touch, and hear.*

Music Makers

If you could make music with your legs, wouldn't you do it all the time? Grasshoppers can, and they do!

Usually only male grasshoppers can make music. They use their **talent** to identify themselves or to attract female grasshoppers. They rub their wings together, or drag one of their legs across one of their wings to create sound. Some species have their own special songs, so all the other grasshoppers know who is singing.

To hear each other's songs, grasshoppers have developed a good sense of hearing. They use their antennae and other organs on or near their legs to hear one another's music.

The male grasshopper sings to ▶ attract the female grasshopper.

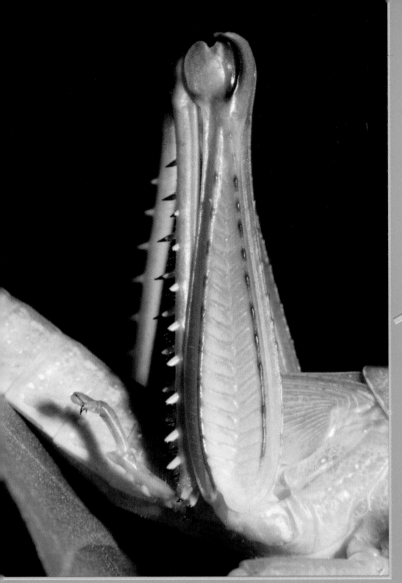

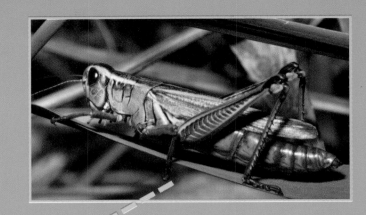

12

Long Jumpers

Wouldn't it be neat if you could jump across an entire football field? It may sound impossible, but if you were a grasshopper, it would be easy. Some grasshoppers can jump more than 100 times their length. They move around by walking, flying, or jumping.

Grasshoppers may be small, but they're very powerful. A grasshopper's body has about 900 **muscles**. That's 200 more muscles than your body has. Their back legs are longer than their entire bodies, and they use the muscles in these legs to **launch** themselves into flight.

◀ *Grasshoppers use the strong muscles in their legs to jump.*

Grasshoppers All Around

Most grasshoppers spend their lives in one home, but some travel from place to place, or **migrate**. Grasshoppers usually only fly a little way at a time, but they may take many trips. These trips can add up to hundreds of miles. Can you imagine traveling 100 miles without a car? It would take a very long time!

Grasshoppers live all over the world, except at the North and South Poles. Some grasshoppers live in trees and on the leaves of plants, and others hang out on the ground in grassy meadows and fields. Some grasshoppers prefer sunny, sandy beaches. Others like cool, dark forests. Where would you live if you were a grasshopper?

Grasshoppers live in many different places. ▶
They even live on the beach!

15

Some grasshoppers eat flowers.

Grasshoppers are prey to beetles.

Prey and Predators

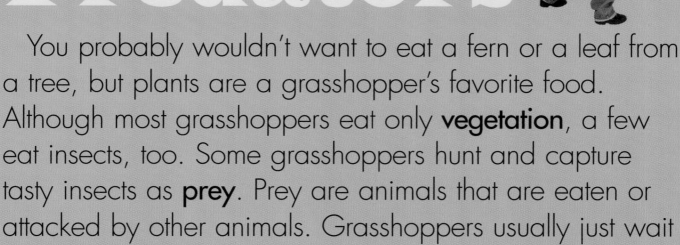

You probably wouldn't want to eat a fern or a leaf from a tree, but plants are a grasshopper's favorite food. Although most grasshoppers eat only **vegetation**, a few eat insects, too. Some grasshoppers hunt and capture tasty insects as **prey**. Prey are animals that are eaten or attacked by other animals. Grasshoppers usually just wait to see if a juicy insect happens by.

Grasshoppers can be prey, too. They have many **predators**. Predators are animals that live by eating or attacking other animals. Beetles, birds, flies, lizards, mice, snakes, spiders, and wasps all find grasshoppers a yummy treat. Life as a grasshopper isn't easy.

◀ *A lizard is often a threat to grasshoppers.*

Blending in and Staying Safe

Have you ever been scared of a bird? If you were a grasshopper, you would be. Birds are some of a grasshopper's worst enemies. Grasshoppers' main **defense** against their predators is **camouflage**. They stay still and try to blend into their surroundings so predators can't see them. Different grasshoppers have different colors to help them camouflage themselves. Grasshoppers that live in grass and on leaves are bright green, grasshoppers on the beach are sand-colored, and grasshoppers on the ground are brown like the dirt.

In case they're spotted by predators, grasshoppers have several other defenses. They use their powerful legs to jump or fly away. They can also use their strong jaws to bite.

Grasshoppers are all different colors so they can camouflage themselves in different environments. ▶

The cone-headed grasshopper is green so it can blend in with the leaves.

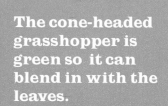

This grasshopper is brown to blend in with the stem.

The Carolina Locust is camouflaged in the dirt.

19

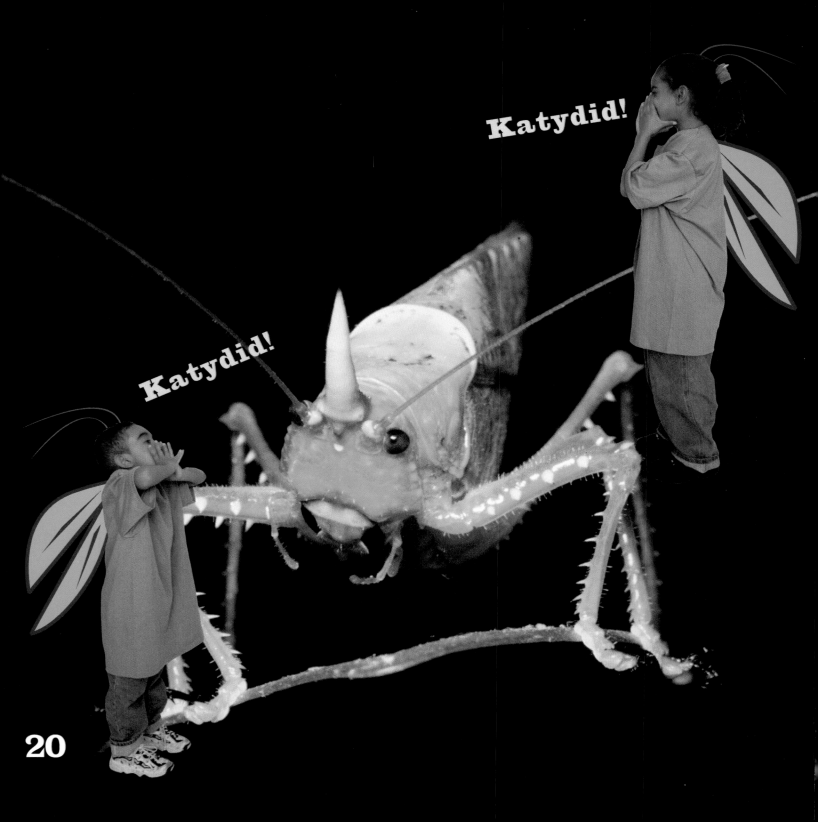

20

"Katydid, Katydid"

When grasshoppers sing, they're calling to one another. However, one special kind of grasshopper may sound as though it's singing to you. That's because it sounds like it's actually speaking in English. This grasshopper is the katydid.

Katydids are large, green- or brown-colored long-horned grasshoppers. They live in trees or on plants that grow near the ground. Katydids get their name from the sound of their song. Male katydids sing to females, calling, "Katydid, katydid." They start singing when the sun goes down and can sing all night long without getting tired. You can hear them most during the late summer and early autumn.

◀ *Katydids can sing all night long.*

Swarm!

   If your whole town or city picked up and started walking to a new place, you'd be a pretty big group. Sometimes grasshoppers **swarm**, or fly in huge groups to new locations. One swarm of grasshoppers was over 2,000 miles long. That's almost as long as the distance from one end of the United States to the other.

   Though grasshoppers that swarm can sometimes be dangerous, most grasshoppers are harmless. They sing to each other, relax in the grass or on the beach, look for food, and try to stay safe. It might be nice to be a grasshopper!

Glossary

antennae (an-TEH-nee) Feelers located on an insect's head.

camouflage (KA-muh-flahj) To hide by blending into one's surroundings.

classify (KLA-sih-fy) To organize or group objects.

defense (duh-FENS) A way to protect one's self.

insects (IN-sekts) Group of animals (also called bugs) that have several key things in common.

launch (LAWNCH) To push forth into the air.

migrate (MY-grayt) To travel from one place to another place.

muscle (MUH-suhl) A body part that helps animals to move.

predator (PREH-duh-ter) Animals that live by eating or attacking other animals.

prey (PRAY) Animals that are eaten or attacked by other animals.

swarm (SWORM) A large group of insects traveling together.

talent (TAL-unt) The ability to do something very well.

vegetation (veh-jeh-TAY-shun) Plants.

Index

Web Sites:

You can learn more about grasshoppers on the Internet.
Check out this Web site:
http://www.insect-world.com/main/orthopta.html

HOGANSVILLE PUBLIC LIBRARY